Seed Production and Technology

CONTENT VIBES
PUBLISHERS · PRINTERS · AGGREGATORS

About the Authors

Dr. S. Prabakaran has done his B.Sc. (Agri.) from TNAU sub-campus at Trichy. He studied his M.Sc. and Ph.D. from Indian Agricultural Research Institute (IARI), New Delhi. His area of specialisation was Plant Genetic Resources. In 2015, he received ICAR-JRF 22nd rank in plant sciences and secured merit fellowship. In 2017, he secured 1st rank IARI entrance exam and secured IARI merit fellowship for Ph.D. program. He also qualified Agricultural Scientist Recruitment Board (ASRB) – National Eligibility Test (NET) in 2017. He has qualified UGC-SRF and lectureship in Environmental sciences in 2019. Currently, he is teaching agriculture optional to civil service aspirants at Chennai.

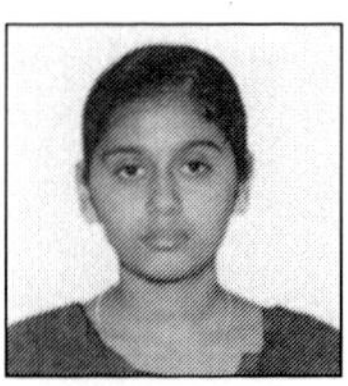

R. Sowmyapriya have studied her B.Sc. (Agri.) from TNAU sub-campus at killikulam, Tamil Nadu. She studied her M.Sc. (Biotechnology) from UAS, Bangalore. Currently, she is pursuing Ph.D. (Biotechnology) at IARI, New Delhi. She was the recipient of JNU fellowship and DBT fellowship.

A. Krishnamoorthi I did my B.Sc. agriculture (2012-16) from VIA Agriculutre college Affiliated to TNAU and I did my Msc Agriculture in Plant geneticr esource (2108-20) from TNAU and I am written 10 book chapters 2 research article and 10 review articles I cleared my ICAR SRF Examination with Rank 4 in the Discipline of Economic Botany and Plant Genetic Resources. Now I am pursuing my Ph.D. at the ICAR-National Bureau of Plant Genetic Resourrces IARI Pusa Campus New Delhi.

Seed Production and Technology

S. Prabakaran
R. Sowmyapriya
A. Krishnamoorthi

CONTENT VIBES
PUBLISHERS · PRINTERS · AGGREGATORS

CONTENT VIBES
PUBLISHERS · PRINTERS · AGGREGATORS

1168, Sector 13, Urban Estate
Karnal-132 001, Haryana
Tel: 91-84470 75807, 18440 41168
Email: contentvibesppa@gmail.com
www.contentvibes.in

Print ISBN: 978-93-6134-816-7
e-book ISBN: 978-93-6134-995-9

Preface

Seeds are the foundation of agriculture, representing the beginning of the food production chain and the cornerstone of food security worldwide. "Seed Production and Technology" delves into the intricacies of this critical sector, exploring the science, innovations, and best practices that drive the production of high-quality seeds.

This book aims to provide a comprehensive understanding of seed biology, the latest advancements in breeding techniques, and the technological innovations that enhance seed quality and performance. It addresses the challenges faced by seed producers, including climate change, pest resistance, and the need for sustainable practices, while highlighting solutions and strategies that ensure robust seed systems.

Readers will gain insights into traditional and modern seed production methods, the role of biotechnology and genetic engineering, and the importance of regulatory frameworks in maintaining seed quality. The book also discusses the economic and social impacts of seed technology, emphasizing the importance of collaboration between scientists, farmers, and policymakers.

"Seed Production and Technology" is designed for students, researchers, practitioners, and anyone with a keen interest in agriculture and food security. It serves as a valuable resource for understanding how high-quality seeds are developed, produced, and brought to market, ultimately contributing to the global effort to feed an ever-growing population sustainably.

We hope this book inspires innovation, fosters knowledge, and contributes to the advancement of seed science and technology, ensuring a secure and sustainable agricultural future.

Authors

Contents

1

Seed Production and Processing Technologies

Previous Year Questions

1. What is seed treatment?
2. Factors affecting the seed quality of a crop during seed production?
3. Seed treatment campaign – Role & importance?
4. Agronomic and genetic principles of quality seed production?
5. Varietal seed production and hybrid seed production?

Seed Treatment

Seed treatment refers to the application of fungicide, insecticide or both to the seeds to disinfect(deep seated) and disinfest (over seed coat) them from seed borne or soil borne pathogenic organisms and storage insects. It also refers to subjecting the seed to solar energy exposure or immersion in conditional water.

Types of Seed Treatment

1. **Seed disinfection:** It refers to eradication of fungal spores present within the seed coat or more deep seated tissues. For effective control the fungicide must penetrate into the seed to kill the fungus.
2. **Seed disinfestations:** It refers to the destruction of surface borne organisms that contaminated the seed surface but not infected the seed. Chemical dips, soaks, fungicides applied as dust, slurry or liquids have been found successful.
3. **Seed protection:** To protect the seed and young seedling from organisms in the soil which might otherwise cause delay of the seed before germination.

Chemicals used for Seed Treatment

A) Mercurial Compounds

1. **Organo mercurials** – used for small grains, flax cotton, and safflower. Proper dosage is critical over dosage results in seed injuring and under dosage fails to control the disease.

   Eg: Phenyl Mercuric acetate (PMA) Methoxy ethyl Mercury chloride (MEMC) Ethyl Mercuric chloride (EMC)

2. **Inorganic mercurials**: Are limited to mercuric chloride, mercurious chloride and mercuric oxide. These materials are used to treat the seeds, roots, tubers and vegetable crops.

B) Non Mercurials

1. **Organic Non Mercurials** : such as thiram and captan are widely used. They are less effective than the organic mercurials less damaging to the seeds and less dangerous to the persons handling the seeds. The se fungicides act as seed disinfestants and or seed protectants. Over dosage is not harmful and viability is not effected.

   Eg: Thiram, Captan, Carbendazim, Metalaxyl.

2. **Inorganic Non Mercurials**: Copper carbonate, Copper sulphate, Cuprous oxide are the major inorganic Non Mercuric compounds used as fungicides. Copper carbonate and Copper sulphate are used on wheat for prevention of bunt diseases. Cuprous oxide prevents seed decay and damping of in vegetables.

Role of Quality seeds in success of seed production programmes

- Seed is the first input of agricultural production on which the performance and efficacy of other inputs depend.
- Good quality of seeds can contribute upto 30% increase in productivity.
- "Good seed harvests good crop", a good seed means a seed lot that adheres to all the parameters of minimum seed standard; this seed is generally termed as quality seed.
- A good quality rice seed should be pure, full and uniform in size, free from weeds, insect, disease and other inert matters and more over it should be viable (>80% germination).
- Timely availability of good quality seed as per the requirement plays a major role in the higher grain production of a nation.
- In India, 75% small and marginal farmers are lagging behind in agriculture due to unavailability of resources or inputs including seed.

- Therefore, a strong and vibrant seed production and supply system is indispensable for food security of the country and accelerating growth in agriculture.
- Seed is the highest prioritized input in agriculture, on which agriculture sustains.

Seed Treatment Campaign – Role & importance

- The concept of seed treatment involves the use of biological or chemical agents to control or contain primary soil and seed borne infestation of insects and diseases which affects crop productivity and crop production.
- Presently, 70% requirement of seed is met from the farmers own stock which is used for sowing without seed treatment.
- Seed Treatment plays an important role in protecting the seeds and seedlings from seed and soil borne diseases and insect pests affecting crop emergence and its growth.
- However, many farmers in the country are either not familiar with the practice or do not follow it. The practice of seed treatment is also IPM compatible.
- The adoption of this practice by the farmers across the country requires effective extension strategies making the appropriate chemical pesticides/ bio-pesticides and equipments available to the farmers at their door steps beside making them aware of methods of seed treatment, post treatment handling of seeds and planting materials.
- In view of this, Government of India has decided to launch a countrywide campaign for ensuring 100% seed treatment in all important crops.
- Pesticide industry associations, ATMAs, CIPMCs, KVKs, farmers clubs, SAUs, NGOs, etc. can play an important role in the campaign for 100 percent seed treatment.
- In view of this there is needed to initiate the Seed Treatment Campaign on war footing. This campaign is expected to ensure that no seeds of major crops be sown without seed treatment by the farmers.
- Seed distribution/ selling agencies should not sell the seeds without treatment. Campaign coordination and monitoring committees are to be constituted at State/District/ Block level by the Commissioners/Directors of the State Agriculture Departments to gear up the efforts by various agencies involved in the programme.
- Awareness creation material is required to be prepared and distributed among the farmers by various agencies involved in the programme.

- States/UTs are advised to identify the major kharif crops sown in their states, quantity of seed required for each crop and quantity to be treated at farmers level (excluding the seed sold by companies with seed treatment), quantity of chemical pesticides (insecticides & fungicides) and bio-pesticides required and their availability , number of seed dressing equipment required and their availability etc. and plan for the Campaign.
- This is only an indicative, and each State is expected to modify to suit the local conditions/crops within the recommended practices.
- States are so requested to launch the Campaign and ensure promotion of seed treatment practice. National Kharif Conference, held on March, 2013 has also endorsed this Campaign.

Benefits of Seed Treatment

1. Prevention of spread of plant diseases both systemic and non-systemic. Seed treatments is effective in controlling systemic diseases like smut of wheat, Helminthosporium blight of barley, loose and covered smuts of oats etc. Non systemic diseases that infect seed during harvest or storage period such as Fusorium with blight of barley oats, rice, sorghum etc can be effectively controlled by appropriate seed treatment.
2. Seed treatment protects seed from seed rot and seedling blights. Once the seed is planted the protective coating around the seed, acts as a barrier against seed borne and soil borne organisms.
3. Improves germination – Seed treatment improves the germination through the control of surface moulds and flora, which are not pathogenic but may infect the seed during moist harvesting and storage condition. In the germination test thus may kill or cover the seed be fore it has germinated.
4. Provides protection from storage insects and pests. For complete protection it is necessary to treat the seeds with insecticide also.

Factors Affecting the Seed Quality of A Crop During Seed Production

Pre-harvest factors

1. **Harvest Maturity:** Harvest maturity refers to the stage of development at which seeds are harvested. It directly affects their quality. Harvesting too early may result in immature produce with less flavour and reduced nutritional content. Conversely, harvesting too late can lead to overripeness, causing the produce to become soft, less palatable, and more susceptible to decay.

2. **Climatic Conditions:** Weather and environmental conditions during the growing season play a vital role in determining the maturity and quality of seed. Adequate sunlight, temperature, and water availability promote proper growth and development, leading to better-quality produce. Extreme climatic conditions, such as excessive heat or cold, can affect the maturity process and cause irregularities in the produce. Pre-harvest rain may also affect the viability.
3. **Cultural Practices:** Farming practices, including irrigation, fertilization, and pest control, have a direct impact on seed quality. Providing adequate water and nutrients helps plants grow optimally and produce high-quality seeds. Effective pest management ensures that the produce is free from damage caused by pests and diseases.
4. **Pathogen and Disease Management:** Effective pre-harvest disease management is crucial for post-harvest quality. Fungal and bacterial infections during the growing season can lead to post-harvest decay and spoilage. Proper disease management practices help reduce post-harvest losses and maintain product quality.
5. **Genetic factors:** The storage is influenced by the genetic make up of the seed. Some kinds are naturally short lived (e.g) onion, soybeans, ground nut etc., Based on the genetic make up seeds are classified into

   Micro biotic – short lived

   Meso biotic- medium lived

   Macro biotic – long lived
6. **Provenance:** The seeds harvested in different climates (or) at different times show differences in viability. Because they would have been subjected to different pre harvest conditions which will have caused different amounts of deterioration by the time, the seeds are harvested.

Post-harvest Factors

1. Seed moisture content

The amount of moisture in the seeds is the most important factor influencing seed viability during storage. Generally if the seed moisture content increases storage life decreases. If seeds are kept at high moisture content the losses could be very rapid due to mould growth. Very low moisture content below 4% may also damage seeds due to extreme desiccation or cause hard seededness in some crops.

For cereals in ordinary storage conditions for 12-18 months, seed drying up to 10% moisture content appears quite satisfactory. However, for storage in sealed containers, drying upto 5-8 % moisture content depending upon particular kind may be necessary.

Harringtons thumb rule on seed moisture content : For every one per cent decrease in seed moisture content the life of seed will be doubled. This is again hold good between 4- 12°C. Based on the tolerance and susceptibility of seeds towards moisture loss seeds are classified into

Orthodox – the seeds able to tolerate moisture loss and less seed moisture favours the storage. i.e. decreased moisture increased storage period. Eg. Rice, sorghum, and most of the cultivated species.

Recalcitrant – just opposite to the orthodox. Seeds not able to tolerate moisture loss. Required high moisture for viability maintenance.

2. Storage pest

Fumigation to control insects will also help in longer period of stroage. Fumigants (e.g) methyl bromide, hydrogen cyanide, ethyline dichloride, carbon tetra chloride, carbon disulphide and napthalene and aluminimum phosphine.

3. Temperature

Insects and moulds increase as temperature increases. The higher the moisture content of the seeds the more they are adversely affected by temperature. Decreasing temperature and seed moisture is an effective means of maintaining seed quality in storage. The following thumb rules by Harrington are useful measures for assessing the effect of moisture and temperature on seed storage. These rules are as follows.

1. For every decrease of 1% seed moisture content the life of the seed doubles. This rule is applicable between moisture content of 5-14%.
2. For every decrease of 5°C in storage temperature the life of the seed doubles. This rule applies between 0°C to 50°C.
3. Good seed storage is achieved when the % of relative humidity in storage environment and the storage temperature in degrees Fahrenheit add upto one hundred but the contribution from temperature should not exceed 50 °F

4. Relative Humidity

Relative Humidity and temperature are the most important factors determining the quality of seeds during storage. Seeds attain a specific and characteristic moisture content when subjected to given levels of atmospheric humidities. This characteristic moisture content called equilibrium moisture content.

Equilibrium moisture content for a particular kind of seed at a given Relative Humidity tends to increase as temperature decreases. Thus the maintenance of seed moisture content during storage is a function of relative humidity and to a lesser extent of temperature. At equilibrium moisture content there is no net gain or loss in seed moisture content.

5. Gas During Storage

Increase in O_2 pressure decrease the period of viability. N_2 and CO_2 atmosphere will increase the storage life of seeds.

Agronomic and genetic principles of quality seed production

Genetic principles

1. **Developmental Variation:** When a seed crop is grown in difficult environmental conditions such as different soil and fertility conditions, under saline or alkaline conditions or under different photo-periods or different elevations or different stress conditions for several consecutive generations the developmental variations may arise as differential growth response.
2. **Mechanical Mixtures:** This is the major source of contamination of the variety during seed production. Mechanical mixtures may take place right from sowing to harvesting and processing in different ways such as;
    a. Contamination through field – self sown seed or volunteer plants
    b. Seed drill – if same seed drill is used for sowing 2 or 3 varieties
    c. Carrying 2 different varieties adjacent to each other.
    d. Growing 2 different varieties adjacent to each other.
    e. Threshing floor
3. **Mutations:** It is not of much importance as the occurrence of spontaneous mutations is very low i.e. 10-7. If any visible mutations are observed they should be removed by rouging. In case of vegetatively propagated crops periodic increase of true to type stock would eliminate the mutants
4. **Natural Crossing:** It is an important source of contamination in sexually propagated crops due to introgression of genes from unrelated stocks/ genotypes. The extent of contamination depends upon the amount of natural cross-fertilization, which is due to natural crossing with undesirable types, offtypes, and diseased plants.
5. **Genetic drift:** When seed is multiplied in large areas only small quantities of seed is taken and preserved for the next years sowing.

Because of such sub-sampling all the genotypes will not be represented in the next generation and leads to change in genetic composition.

6. **Minor Genetic variation:** It is not of much importance, however some minor genetic changes may occur during production cycles due to difference in environment. Due to these changes the yields may be affected.
7. **Techniques of the Breeder:** Instability may occur in a variety due to genetic irregulaeities if it is not properly assessed at the time of release. Premature release of a variety, which has been breed for particular disease, leads to the production of resistant and susceptible plants which may be an important cause of deterioration. When sonalika and kalyansona wheat varieties were released in India for commercial cultivation the genetic variability in both the varieties was still in flowing stage and several secondary selections were made by the breeders.
8. **Breakdown of male sterility:** Generally in hybrid seed production if there is any breakdown of male sterility in may lead to a mixture of F1 hybrids and selfers.

Agronomic Principles

1. **Selection of the Agro-Climatic Region:** A crop variety employed for seed production in a given area must be adapted to the photoperiod and temperature conditions prevailing in that area. Regions of moderate rainfall and humidity are more suited for seed production than the regions with high rainfall and humidity.
2. **Field Preparation:** Field preparation consists of eliminating any weeds and volunteer plants and making flat or raised seedbeds, as required. Well tilled seed bed help in improved germination, good stand establishment and removal of potential weeds during tillage operations.

   Stale seedbed technique, in which the seedbeds are formed about a week before it is to be sown, can be adopted. This allows the weed seeds to germinate first, so these can be removed to minimize the weed competition in the seed bed before the seed crop is sown.
3. **Isolation**: Proper isolation distance should be provided to avoid natural cross pollination and spread of diseases. The isolation distance varies from crop to crop and depends on nature of contamination and direction of the prevailing wind. Generally more isolation is required at this stage than the later stages. Distance or time isolation can be practiced to avoid contamination.

4. **Planting ratio** : The row ratio of female and male parental varies from region to region depending on weather conditions and potentiality of parental lines. The commonly adopted planting ratios of male and female are 2:8, 2:6 or 3: 8.
5. **Synchronization of flowering**: When both A and R-line are of same duration sowing of A-line should be adjusted with second sowing of R-line. If A and R lines are of different growth duration, the difference in duration should be adjusted with second sowing of R-line. For example if A-line comes to flowering in 65 days and R-line in 72 days then the difference is 7 days. After second sowing of R-line adjust the sowing of A-line with a gap of 7 days I.e. if First sowing of R-line is done on 1st June, Second sowing on 5th June and third sowing on 10th June, then sowing of A-line should be done on 12th June)
6. **Roguing:** It is the selective removal of undesirable plants from a seed crop on the basis of distinct morphological characteristics inorder to improve one or more parameters (genetic purity, free from diseases and noxious weeds) of seed quality. Rogues include off-types, diseased plants, objectionable weeds and other crop plants.
7. **Weed control:** Weed plants in the seed field or near by area may also serve as the host to a number of diseases. Presence of prohibited weed (objectionable/obnoxious weed) seed may result in the rejection of an entire seed plot/lot. For instance, by the presence of more than 1 and 2 wild rice plants (Oryza sativaL.var.fatua)in a population of 10,000 plants of rice (O.sativa) crop in foundation & certified seed plots, respectively, will be rejected.

Varietal seed production and hybrid seed production

Varietal seed production	Hybrid seed production
It is single parent multiplication	It needs two to many parents
Isolation distance requirement is less	Isolation distance requirement is more
Production is by open pollination	Production is by managed control pollination (Female)
Seed can be used continuously for 3/4/5 generations	Seed has to be changed every time
Production technique is uniform (multi-plication)	Technique differ with crop
Production care is less	Production care is more
Yield will be lower	Yield will be higher
Profit is less	Profit is higher

2

Seed Certification Seed Testing and Storage

Previous Year Questions

1. Different Classes of Seeds?
2. Various parameters of field & seed standards for breeder seed production?
3. Factors that affect the longevity of seeds during storage in warehouses?
4. Initial Seed Quality?
5. Seed Testing?

Different Classes of Seeds

1. **Nucleus Seed:** It is handful of seed maintained by concerned breeder for further multiplication. The nucleus seed will have all the characters that the breeder has placed in it and it is of highest genetic purity. The quantity of nucleus seed is in kilograms.
2. **Breeder Seed (BS):** Breeder seed is a seed or vegetative propagating material directly controlled by the originating or sponsoring plant breeder of the breeding programme or institution and or seed whose production is personally supervised by a qualified plant breeder and which provides the source for the initial and recurring increase of foundation seed. Breeder seed is exempted from certification. However, the breeder seed crop is being inspected at appropriate stages by a Monitoring Team consisting of breeder, representative of NSC and representative of Seed Certification Agency. The breeder seed lots are affixed with Golden Yellow tags. The genetic purity of breeder seed crop should be maintained at 100 per cent.
3. **Foundation Seed (FS):** It is the progeny of breeder seed, or be produced from foundation seed, which can be clearly traced to breeder seed. The production is supervised and approved by Certification Agency and so handled to maintain genetic purity and physical purity and required to conform to certification standards. Thus, foundation seed (Stage II)

can even be produced from foundation seed (Stage I). Indian Minimum Seed Certification Standards (IMSCS) are same for Stage I and II. The foundation seed lots are affixed with white bags. The genetic purity of foundation seed should be maintained at 99.5 per cent.

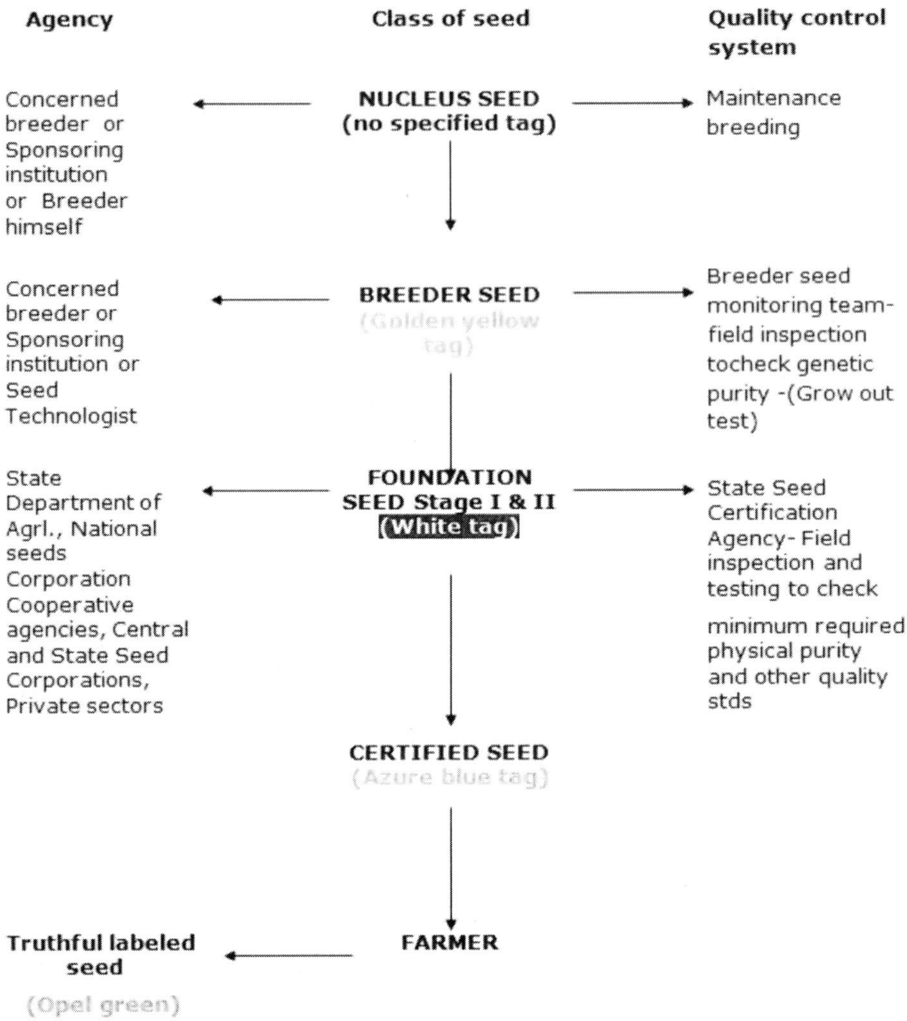

4. **Registered seed:** Registered seed shall be the progeny of foundation seed that is so handled as to maintain its genetic identity and purity according to standard specified for the particular crop being certified. A purple colour certificate is issued for this category of seed. The registered seed is generally avoided in India & certified seed is produced directly from foundation seed.

5. **Certified Seed (CS):** Certified class seed shall be the progeny of foundation seed and its production shall be so handled to maintain specific genetic identity and purity according to standards prescribed for the crop being certified. Certified seed (Stage III) may also be the progeny of certified seed (Stage I). Such multiplication should be done after getting permission from the Certification Agency. The certification agency shall give such permission provided the genetic purity will not be significantly altered. The certified seed lots are affixed with azure blue tags. This is the commercial seed which is available to the farmers and its genetic purity should be 99 per cent.
6. **Truthfully labeled (TFL) seeds:** Whereas truthfully labeled seed is one which is being produced and marketed by the producing company by maintaining the labeling standards. The farmer or the user of the seed does not know the pedigree of the truthfully labeled seed and he has to relay on the seed producing company. Labeling is compulsory but certification is voluntary.

Essential Requirements for Production of Certified Seeds

1. Source of Seed

Appropriate / proper class of seed need to be obtained from approved source according to stages of seed multiplication. For certified seed production, certified foundation seed with white tag is to be used.

2. Registration of Seed Plot

When the seed plot is to be offered for certification, it is necessary to register the said plot for certification with District Seed Certification Officer. For this an application is to be made in prescribed form (Form A) and is to be submitted to District Seed Certification Officer along with agreement bond on stamp paper of Rs.100/ duly notarised or registered with Taluka magistrate. The registration fee is to be paid in the treasury on challans as per following rates. The Application form with agreement bond are to be handed to seed certification office along with a copy of challans.

S.No.	Stage	Type	Fees to be Paid
1	Certified	Varieties	Rs. 80.00 per acre seed plot area
2	Certified	Hybrids	Rs. 160.00 per acre seed plot area

3. Land Requirement

The land selected for seed production should be comparatively free from soil borne diseases, insect pests and noxious weed seeds. As far as possible the land selected for seed production should not have same crop grown in the preceding season. However, if same crop is grown in the preceding season, then it should be irrigated three weeks before sowing so that seed of previous crop that had fallen the soil will start germination. The land should be harrowed for removing seedlings of volunteer plants, weeds, other crop plants.

4. Isolation Requirement

Isolation is the separation of seed field from the fields of

a) Other varieties or hybrids of same crop.
b) Same variety of hybrids which do not conforming to varietal requirements.
c) Other related species which are readily crossable and
d) Fields affected by designated diseases prevent genetic and disease contamination.

Isolation of seed plot can be maintained by two ways i.e.

a) Time isolation and b) Space isolation.

a) Time isolation: Sowing of seed plot is adjusted in such a manner that the seed plot does not came to flowering at the same time with the neighbouring crop of same variety or other variety of the same crop or related species. Sowing of the seed plot is usually carried 15 to 21 days before or after emergence of the neighbouring crop. This time isolation helps is preventing the genetic contamination of the seed crop as seed plot comes to flowering either before or after the neighbouring crop completes its flowering.

b) Space Isolation: Space isolation is the minimum distance kept between the seed plot and neighbouring plot of same crop which prevents natural cross pollination and physical contamination.

Rice (Certified seed production) – 3m

Sunflower (Certified seed production) – 200m

5. Roguing

Rouging consists of removal of

a) Off types
b) Volunteer plants
c) Pollen shedders in female (A) lines,
d) Plants of noxious weeds and other crops,
e) Diseased plants affected by seed borne diseases growing in the seed plot and

f) Tassels from plants in the female rows of seed production of single hybrids and double hybrids in maize.

6. Field Inspection

As per provisions of seed certification, the seed plots offered for certifications are subjected to field inspection by the staff of seed certification agency. The number of field inspections is usually carried without prior intimation to the seed producer. It is the responsibility of the seed producer to follow the instructions given by the field inspector. For this purpose seed producer or his responsible representative remain present on the field during each inspection. During the field inspection, source of seed used with tag isolation planting rouging are checked and counts of off types, shedders etc are taken.

7. Harvesting, Threshing, Drying and Sealing of Raw Seed

Seed plot should be harvested at proper stage of maturity and only after permission is granted by the field inspector. The seed produce be threshed and winnowed for removing major part of inert matter. i. e Stones, sand , dried twigs, leaves, husk etc. The cleaned seed produce should be bagged in the presence of the field inspectors who will seal the entire bags ad issue threshing certificate. At the time of sealing, field inspector draw 3 kg sample from each lot. The seed bags should be transported to authorized seed processing plant for processing along with threshing certificate. It may be noted that the seed will not be accepted for processing at seed processing plant unless it is brought in bags sealed by the field inspector and issued with threshing certificate.

8. Seed Processing

Seed lot accepted for processing is processed at the seed draws three samples and sends one sample to seed testing laboratory for testing laboratory for testing, one sample kept at his disposal and one sample for concerned seed producer.

9. Bagging, Tagging

When the seed lot is passed by seed certification agency on the basis of seed testing laboratory report (STL). The processed and treated seed is bagged and tagged with appropriate tags issued by seed certification officer.

Various steps involved in seed certification

It can be completed in six broad phases

1. Receipt and scrutiny of the application.
2. Verification of seed source, class and other requirements.

3. Field inspection should be conducted to see that fields are up to the prescribed field standard.
4. Post harvest inspection, including processing and packing.
5. Seed sampling and testing to confirm that the seeds are up to the prescribed seed standards.
6. Grant of certificate, tagging and sealing.

1. **Receipt and scrutiny of the application:** All those persons who are interested in seed certification should submit an application in Form No 1 to the concerned seed certification officer with the prescribed fees of Rs 25/-. The fee is for one season for a single variety and for an area up to 25 acres (10 ha.) If the area is more than 25 acres or if more than one variety is planted separate applications should be made for each variety. If the area is less than 25 acres under one variety but if the fields are scattered and separated by more than 50 meters separate applications should be made. On receiving the applications the seed certification agency verifies for the following conditions:

1. **Eligibility of the variety:** Only those varieties that are notified by the central govt. are eligible for certification.
2. **Establishing the seed source:** The seed producer should submit the tag, invoice, and a copy of Form No. 2).
3. There should not be any difficulty in reaching the field for carrying out timely field inspection.
4. Whether the required isolation and land requirement is followed or not.
5. Whether the processing plant facility is available to the applicant.
6. Whether the applicant has paid the requisite registration fee or not.

If all the six conditions are fulfilled then the seed producer has to pay the field inspection fees as given below:

Various Certification Charges

1. Cost of the form No 1 : Rs 2.00
2. Registration fee (per unit) : Rs 25.00
3. Inspection fee (per ha.)
   a. Self-pollinated Crops : Rs 125.00
   b. Cross Pollinated Crops : Rs 175.00
   c. Other than Cotton hybrids/parents : Rs 175.00
   d. Cotton Hybrids : Rs 800.00
   e. Vegetable Crops : Rs 150.00

4. Grow Out Test (per sample) : Rs 150.00
5. Seed Testing
   a. Routine tests : Rs 30.00 per sample
   b. Health tests : Rs 5.00 per sample
   c. Revalidation Charges (sample) : Rs 30.00

2. **Verification of seed source, class and other requirements:** The seed should be from authentic source and from appropriate class and should be in accordance with Indian Minimum Seed Certification Standards.
3. **Inspection of Seed Fields**: The certified seed producers should grow and harvest the crop as per the guidelines issued by the seed certification agency. They must carefully and faithfully carry out the roguing and other operations as per the directive of the certification agency. The certification staff conducts field inspections at appropriate stages of crop growth to ensure that minimum standards of isolation, preceding crop requirement, roguing and other special operations are maintained at all times. The inspection of seed crop is done at different stages of crop growth such as at the time of sowing (when new crop is introduced), vegetative stage or preflowering stage, flowering stage, post flowering or preharvest stages and at the time of harvest. The contaminants to be observed during field inspections are offtypes, pollen shedders, shedding tassels, inseperable other crop plants, objectionable weed plants and diseased plants.

Method of Taking Field Counts

1. Determine the number of field counts. For all crops a minimum of five counts are to be taken for an area up to two hectares, and an additional count is to be taken for each additional two hectares or part thereof as given below.

Area of the field in hectares	Minimum number of counts to be taken
Upto 2	5
2-4	6
4-6	7
6-8	8
8-0	9
Above 10	10

In any inspection, if the first set of counts show that the seed crop does not confirm to the prescribed standards for any factor, a second set of counts should be taken for that factor, if the percentage of first set of count for that factor is not more than twice the maxmimum permissible level. Two sets of counts are called as double counts. In hybrid seed production plots, the number of counts must be taken separately for both the parents.

2. Number of plants to be observed for completing one count. The number of plants to be observed for completing a single count varies from crop to crop. The number of plants/heads to be observed for completing a single count is given below.

Crop	Number of plants / heads per count
Wide spaced crops: Bhendi, Brinjal, Bulb crops, Chillies, Cole crops, Cotton, Cucurbits, Groundnut, Maize, Potato, Redgram, Tomato, Root crops, etc. **Medium spaced crops:** Beans, Cowpea, Gram, leaf crops, Moong, Urad, Mustard, Peas, Sesame, Sunnhemp, etc.	100 plants 500 plants
Thickly sown crops: Berseem, Jute, Lucern, Mesta, Soybean, Bajra, Paddy, Wheat, Sorghum, etc.	 1000 plants 1000 heads

3. **Taking of Filed Counts:** The procedure for taking filed counts differs for different crops.
4. **Rejection of seed fields:** All the seed fields, which do not confirm to the required standards for any of the factors should be rejected. The rejection letter should be immediately communicated to the seed grower stating the reasons for the rejection.
5. **Post Harvest Inspection:** The personnel from the seed certification agency should inspect the fields during harvesting or post harvesting, so that there are no mechanical mixtures and the seed is not handled badly during threshing or afterwards. Then the seed is sent to seed processing plant with a threshing certificate. The personnel from the seed certification agency will be inspecting the seed processing plant to avoid mechanical mixtures and damage caused to the seed during processing.
6. **Seed Sampling and Testing:** The representative from seed certification agency draws a representative sample from the seed lot at the time of processing or after processing and sends the sample to official seed testing laboratory for evaluation. In the seed testing laboratory the samples will

be evaluated for seed standards such as pure seed, inert matter, other crop seed, weed seeds, germination percentage and moisture percentage etc.

7. **Grant of certificate, tagging and sealing:** After receiving a satisfactory report from the seed testing laboratory, tagging and sealing of bags will be done under the supervision of seed certification agency. Under special circumstances, advance tags will also be issued to the extent of 75 per cent of the seed lot. Tags and seals should be in accordance with general seed certification requirements. Affixing of tags and seals on the containers completes the process of certification of seeds.
8. **Control Plot testing:** The seed certification agency should arrange for a postseason grow-out test for all hybrids as prescribed in the standards. Randomly samples should be drawn from certified seed lots and sent to grow-out test to check the efficiency and accuracy of the work done.
9. **Validity period:** The seed is initially valid for a period of nine months from the date of testing the samples. If the seed is not sold within the stipulated period, it can be revalidated for a period of six months if the seed lot meets the required seed standards. The seed can be revalidated as long as it meets the prescribed seed standards and for each revalidation the validity period will be extended for six months.
10. **Revocation of certificate:** If the certification agency is satisfied that the certificate granted by it has been obtained by misrepresentation of essential facts, or the holder of the certificate has failed to comply with the conditions subject to which the certificate has been issued, can revoke the certificate. The certificate can be revoked only after giving a show cause notice to the holder of the certificate.
11. **Appeal against seed certification agency:** If any certified seed grower is not satisfied by the decision taken by the seed certification agency (in rejecting the seed plot), he can make an appeal to the appellate authority specified by the state government. The appeal should be made within 30 days from receiving the rejection letter. The application should be submitted personally or it should be sent through registered post. The decision of the appellate authority will be final and it is binding on the seed certification agency and the seed grower.

Institutions Associated with the Production of Certified Seeds in India

On 29th December 1966, the Seeds Act was passed. It came into force throughout the country on 2nd October 1969.

The main features of the Seeds Act, 1966 are discussed below:

1. **Applicability:** It is applicable only to notified kinds/varieties of seed and vegetatively propagating materials used for sowing.
2. **Sanctioning legislation:** The Act provides for the formation of an apex advisory body, namely, the Central Seed Committee; the Central seed Certification Board; establishment of Seed Certification Agencies; and Central and State Seed Testing Laboratories, etc.
3. **Regulatory legislation:** The Act provides for the provisions for notification of kinds/varieties to be brought under the purview of the Seeds Act; regulation regarding the sale of seed; and the establishment of a suitable seed law enforcement machinery. Under the Act the Central Govt. is empowered to make rules to carry out the purposes of the Act and to give directions to State Govts., it necessary, for carrying into execution, in the state concerned, the provisions of the Act or Rules.

Seed Law Enforcement

The responsibility for enforcing various provisions regarding regulation of sale of seeds of notified kinds/varieties rests with the seed inspectors.

The seed inspector is required to

a) Inspect of frequently as may be required all places used for the growing, storage or sale of any seed of any notified kind or variety;

b) Satisfy himself that the conditions of the certificates/labels are being observed;

c) Procure and send for analysis, if necessary, samples of any seeds, which he has reason to suspect, are being produced, stocked or sold or exhibited for sale in contravention of the provisions of the Act of the Rules.

d) Investigate any complaint which may be made to him in writing in respect of any contravention of the provisions of the Act, or the Rules;

e) Maintain a record of all inspections made and action taken by him in the performance of his duties, including the taking of samples and the seizure of stocks and to submit copies of such records to the Director of Agriculture, or the certification agency.

f) When so authorized by the State Government, to detain imported containers which he has reason to suspect contain seeds, the import of which is prohibited, except and in accordance with the provisions of the Act and the Rules;

g) Institute prosecution in respect to breaches of the Act and the Rules; and

h) Perform such other duties as may be entrusted to him by the competent authority from time to time.

Powers of Seed Inspectors

1. To take samples of seed of any notified kind/variety from any person selling such seed, or purchaser or consignee, and to send such samples for analysis to the seed analyst notified for the area;
2. To enter and search, at all reasonable times, with such assistance, if any, as he considers necessary, any place in which he has reason to believe that an offence under this Act has been or is being committed and order in writing the person in possession of any seed in respect of which the offence has been or is being committed, not to dispose of any stock of such seed for a specific period not exceeding thirty days or, unless the alleged offence is such that the defect may be removed by the possessor of the seed, seize the stock of such seed;
3. To examiner any record, register, document or any other material object found in any place mentioned in clause (2) and seize the same, if he has reason to believe that it may furnish evidence of the commission of an offence punishable under this Act; and exercise such other powers as may be necessary for carrying out the purposes of this Act or any Rule made there under.
4. On demand to pay the cost of seed, calculated at the rate at which such seed is usually sold to the public, to the person from whom the same is taken.
5. To break-open any container in which any seed or any notified kind or variety may be contained, or to break-open the door of any premises where any such seed may be kept for sale; Provided that the power to break-open the door shall be exercised only after the owner or any other person in occupation of the premises, if he is present therein, refuses to open the door on being called upon to do so.
6. Where the seed inspector takes any action under clause (a) of sub-section (1), he shall, as far as possible, call not less than two persons to be present at the time when such action is taken, and take their signatures on the memorandum to be prepared in the prescribed form and manner.
7. The provisions of the Code of Criminal Procedure, 1898, shall, so as may be, apply to any search or seizure under this section as they apply to any search or seizure made under the authority of a warrant issued under section 98 of the said code.

Statutory Bodies and Agencies Established in India under the Seeds Act, 1966

1. **Central Seed Committee:** The Central Seed Committee set up under the Act is the main source of advice to the Central Government on the administration of the Act, and any other matter related to seeds. The committee consists of a chairman, two representatives of seed growers, eight representatives of other interests; nominated by the Central Govt., and one representative of the State Governments. **Director (Seeds), in Ministry of Agriculture, Govt. of India**, acts as the Secretary of the Committee.

The main functions of the Central Seed Committee as envisaged in the Act and Rules are:

a) To advise Central and State Governments on all matters related to seeds.

b) To advise Government regarding notification of such kinds/varieties for which it thinks it has become necessary or expedient to regulate the quality of seeds.

c) To advice Government of the minimum limits for germination and purity for those kinds/varieties brought under the purview of the Seeds Act.

d) To recommend the procedure and standards for certification, grow out tests and analysis of seeds.

e) To recommend to the Central Government the suitability of any seed certification agency established in any foreign country for the purposes of this Act.

f) To recommend the rate of fees to be charged for analysis of samples by the Central and State Seed Testing Laboratories, and for certification by certification agencies.

g) To advise the Central and State Governments regarding suitability of seed testing laboratories.

h) To send its recommendations and other pertinent proposals related to the Act to the Central Government.

i) To carry out such other functions as are supplemental, incidental or consequential to any of the functions conferred by the Act or Rules.

2. **Central Seed Certification Board:** The Government of India has also constituted the Central Seed Certification board to deal with all problems related to seed certification and to co-ordinate the work of State Seed Certification Agencies. The membership consists of a Chairman, 13 members representing interests as the govt. thinks fit, 3 Directors of

Research of Agricultural Universities, 4 Directors of Agriculture in a State; nominated by the Central Government besides a nominated secretary.

3. **State Seed Certification Agencies:** The Seed Act, 1966 provides for the establishment of State Seed Certification Agencies by notification in the official Gazette of State Government / Central Government in consultation with the State Government. On recommendation of the Central Seed Committee these agencies have been established in the form of societies having a governing body and an executive wing. The governing body consists of persons representing all interests, namely, the State Government, seed producing agencies, farmers, subject specialists and seed law enforcement agencies. The Agriculture Production Commissioner or Secretaty of State, Department of Agriculture acts as the Chairman of the Statutory Certification Agency. The governing body lays down the broad policy, while the inspections, seed certification and seed analysis is the responsibility of the executive wing. The functions specified for the Seed Certification Agency under the Act are:

a) Certify seeds of any notified kinds or varieties:

b) Outline the procedure for submission of applications and for growing, harvesting, processing, storage and labeling of seeds intended for certification till the end so as to ensure that seeds lots finally approved for certification meet prescribed standards for certification under the Act, or these rules;

c) Maintain a list of recognized breeders of seeds;

d) Verify, upon receipt of an application for certification, that the variety is eligible for certification, that the seed source used for planting was authenticated and the record of purchase is in accordance with these rules and that the fees have been paid;

e) Take sample and inspect seed lots produced under the procedure laid down by the certification agency and have such samples tested to ensure that the seed conforms to the prescribed standards of certification;

f) Inspect seed-processing plants to see that the admixtures of other kinds and varieties are not introduced;

g) Ensure that action at all stages, e.g., field inspection, seed processing plant inspection, analysis of samples taken and issue of certificates (including tags, marks, labels and seals) is taken expeditiously;

h) Undertake educational programmes designed to promote the use of certified seed, including a publication listing certified seed growers and sources of certified seed;

i) Grant certificates (including tags, labels, seals, etc.) in accordance with the provisions of the Act and these rules;

j) Maintain such records as may be necessary to verify that seed fields for the production of certified seed were eligible under the rules;

k) Inspect fields to ensure that the minimum standards for isolation, roguing (where applicable), use of male sterility (where applicable) and similar factors are maintained at all times, as well as to ensure that seed-borne diseases are not present in the field to a greater extent than those provided in the standards for certification.

4. **Central Seed Testing Laboratory:** The Seed Testing Laboratory at the Indian Agricultural Research Institute, New Delhi, has been notified as the Central Seed Testing Laboratory.

The functions assigned to this laboratory at present are:

a) Initiate testing programme in collaboration with the State Seed Laboratories designed to promote uniformity in test results between all seed laboratories in India.

b) Collect data continuously on the quality of seeds found in the market and make this data available to the Committee;

c) Carry out such other functions as may be assigned to it by the Central Government from time to time; and

d) Act as referee laboratory in testing seed samples for achieving uniformity in seed testing. The State Seed Testing Laboratories are required to send five percent samples to the Central Seed Testing Laboratory along with their analysis results.

5. **State Seed Testing Laboratories:** The Act envisages the establishment of State Seed Testing Laboratories in each State by notification in the official Gazette. The function of the State Seed Testing Laboratory is to carry out the seed analysis work of the State in a prescribed manner.

6. **Appellate Authority:** The Act envisages appointment of an appellate authority / authorities through an official notification in the Gazette by the State Governments to look into the grievances of certified seed producers against a seed certification agency; and that of seed traders against seed law enforcement officials.

Various parameters of field & seed standards for breeder seed production

I) Field Standards

1. **Isolation distance (metre):** To overcome the problem of natural crossing isolation distance has to be maintained. Increase in isolation distance decreases the extent of contamination.
2. **Other crop plants (Max. % by number):** Other crop seed shall include seed units of any plant species other than of pure seed grown as crops.
3. **Other varieties (Max. % by number):** includes the plants of same spp. but different variety.
4. **Weed plants (Obnoxious Max. % by number):** Ex. Wild oat in weed field
5. **Plants infected by seed borne diseases (Max.% of infected plants):** number of plant infected by seed borne pathogens.
6. **General condition of crop:** If the field crop is severely damaged or lodged and irregularly flowered or otherwise damaged that makes assessment difficult will be rejected.

II) Seed Standards

1. **Pure seed:** The seeds of kind / species stated by the sender. It includes all botanical varieties of that kind / species. Immature, undersized, shrivelled, diseased or germinated seeds are also pure seeds. It also includes broken seeds, if the size is >1/2 of the original size except in leguminaceae, and cruciferae where the seed coat entirely removed are regarded as inert matter.
2. **Other crop seed:** It refers to the seeds of crops other than the kind being examined.
3. **Weed Seed:** It includes seeds of those species normally recognized as weeds or specified under Seed Act as a noxious weed.
4. **Inert matter:** It includes seed like structures, stem pieces, leaves, sand particles, stone particles, empty glumes, lemmas, paleas, chaff, awns, stalks longer than florets and spikelets.
5. **Germination%:** Seed germination refers to the ability of a seed when planted under normal sowing conditions to give rise to a normal seedling.
6. **Moisture content:** If seeds are kept at high moisture content the losses could be due to mould growth. So the seeds are dried to recommended moisture content.

Field & Seed Standards for Wheat and Maize

2. Wheat

Seed Standard

S. No.	Factor	Standard		
		Breeder	Foundation	Certified
1.	Pure seed (Min. % by wt.)	99.0	97.0	96.0
2.	Inert matter (Max % by wt.)	1.0	2.0	3.0
3.	Other seed (Max. % by wt.) a) Other crop seed (Max. No. in total; whole sample will be tested) b) Total weed seed (Max. No. in total; whole sample will be (tested)	Trace 2/kg 2/kg	1.0 5/kg 8/kg	1.0 10/kg 10/kg
4.	Germination (Min. %)	85.0	80.0	80.0
5.	Moisture content (Max. %)	12.0	12.0	12.0

Field Standard

S. No.	Factors	Standard		
		Breeder	Foundation	Certified
1.	Isolation distance (in meter)	3.0	3.0	3.0
2.	Other crop plants (Max. % by No)	0.0	0.1	0.2
3.	Other varieties (Max. % by No.)	0.0	0.1	0.5
4.	Weed plants (Obnoxious Max. % by No.) a. Lehli b. Wild Oat c. Wild Lentil	0.0	0.03	0.05
5.	Diseases (Infection by seed borne pathogen: Max no of infected plants) Loose smut (plants/ha)	0.0	12.0	25.0
6.	General condition of crop: if the field crop so damaged or lodget that makes crop assessment difficult to judge the trueness of variety and variety and varietal purity will be rejected.			

2. Maize (Zea mays L) Composites, Synthetics and Open Pollinated Varieties

Seed Standard

S. No.	Factors	Standard		
		Breeder	**Foundation**	**Certified**
1.	Pure seed (Min.% by wt.)	98.00	97.00	97.00
2.	Inert Matter (Max % by wt.)	2.00	3.00	3.00
3.	Other seed (Max. % by wt.)			
	a) Other crop seed (Max. No. in total; whole sample will be tested)	2/kg	5/kg	10/kg
	b) Total weed seed (Max. No. in total; whole sample will be tested)	2/kg	5/kg	10/kg
	c) Other distinguishable varirties based on Kernel colour and texture (Max. %)	2/kg	5/kg	10/kg
4.	Germination (Min. %)	80.00	80.00	80.00
5.	Moisture content (Max.%)	12.00	12.00	12.00
6.	Moisture content in vapour proof container (Max.%)	8.00	8.00	8.00

Field Standard

S. No.	Factors	Standard		
		Breeder	**Foundation**	**Certified**
1.	Isolation distance (in meter)	500	400	400
2.	Other crop plants (Max. % by No)	0.10	0.20	0.50
3.	Other varieties (Max. % by No.)	0.10	0.20	0.30
4.	Weed plants (Obnoxious Max. % by No.)	0.10	0.20	0.30
5.	Plants infected by seed borne diseases (Max. % of infected planted plants)			
6.	General condition of crop: if the field crop is severely demaged or lodged and irregularly flowered or otherwise damaged that makes assessment difficult will be rejected.			

N. B.- Obnoxious weeds & seed borne diseases to be specified by competent authority.

4. Maize (Zea mays L) Hybrids

Seed Standard

S. No.	Factors	Standard		
		Breeder	**Foundation**	**Certified**
1.	Isolation distance (in meter)	-	-	98.00
2.	Inert matter (Max. % by wt.)	-	-	2.00
3.	Other seed (Max. % by wt.) a) Other crop seed (Max. No. in total; whole samole will be tested) b) Total weed seed (Max. No. in total; whole sample will be tested) c) Other distinguishable varietied based on Kernel colour and texture (Max. %)	0.10	0.20	5/kg 5/kg 10/kg
4.	Germination (Min. %)	-	-	85.00
5.	Moisture content (Max. %)	-	-	12.00
6.	Moisture content in vapour proof container (Max %)	-	-	8.00

Field Standard

S. No.	Factors	Standard		
		Breeder	**Foundation**	**Certified**
1.	Isolation distance (in meter)	-	-	400
2.	Other crop plants (Max. % by No)	-	-	0.30
3.	Other varieties (Max. % by No.)	-	-	0.20
4.	Weed plants (Obnoxious Max. % by No.)	-	-	0.30
5.	Plants infected by seed borne diseases (Max. % of infected plants)			
6.	General condition of crop: if the field crop is severely demaged or lodged and irregularly flowered or otherwise damaged that makes assessment difficult will be rejected.			

Factors that affect the longevity of seeds during storage in warehouses

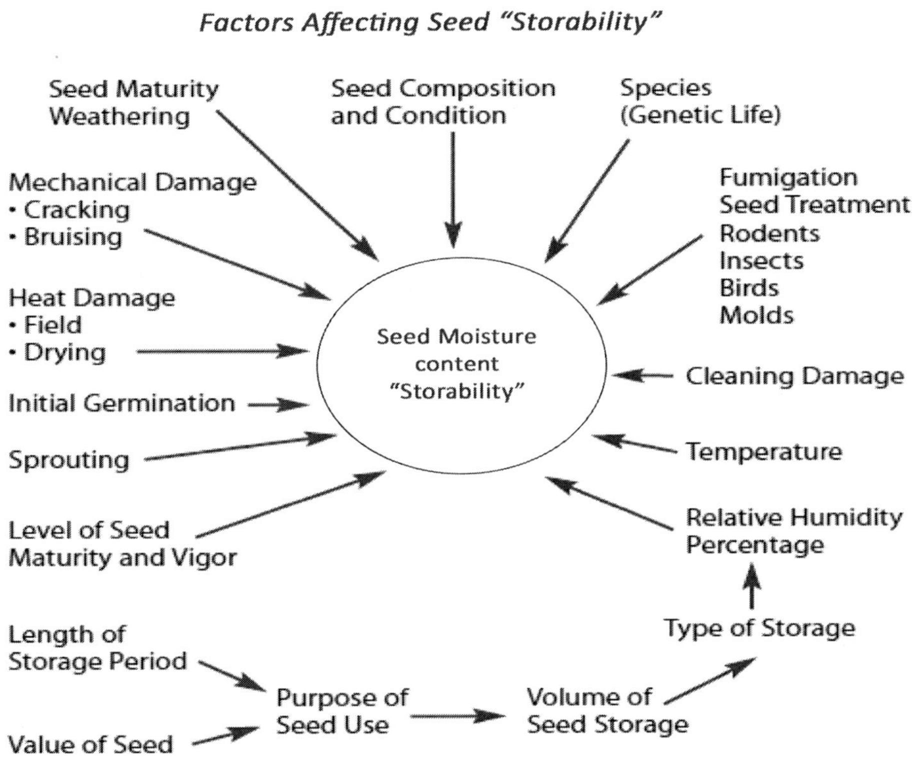

Initial Seed Moisture Content (MC) Initial seed moisture content for storage should range between 5% and 14% to maximize shelf life. When combined with the correct RH, the correct seed MC can help produce safe seed storage conditions for up to many years. The type of seed packaging/container used can also have an effect on shelf life, particularly if the RH is high. Some types of packaging can absorb moisture from the air, or allow it to penetrate to the seed. Seed will then imbibe the moisture, eventually equilibrating with the surrounding air. This can result in shelf life reduction if the seed becomes too moist. RH is not so important if seed is stored in hermetically-sealed containers or other non-absorbent packaging that does not allow the moisture to penetrate in to the seed.

Storage Temperature

Storage temperature must be derived from this formula when assessing storage unit effectiveness: Lower relative humidity means higher allowable storage temperatures, as long as their sum does not exceed a combined value of 100.

For example, storage areas with an RH value of 60% need to be cooled to 40° F to enhance maximum storage potential. Higher temperatures can be especially damaging if the seed is stored at higher moisture contents. Seed stored with lower moisture contents may tolerate higher temperatures better. $°F + RH \leq 100$

Initial Seed Quality

Initial seed quality needs to be established as a base for evaluation of seed deterioration. This can be done as a combination of germination testing and vigor tests (such as cold, emergence rate, EC etc).

Lower temperatures can be facilitated in various ways. Simple solutions include using reflective paint, increasing insulation, or using the more expensive approach of cooling units. With cooling units, the issue of relative humidity must be addressed.

Adequate ventilation can be obtained with the use of various types of fans in combination with correct venting for the room involved.

Moisture exclusion can be maintained with the installation of dehumidification systems. However, this approach can certainly increase the cost of a storage facility. The two methods normally used for dehumidification involve using either chemical desiccants or refrigeration units. Moisture removed through refrigeration/ heating units must exit the storage area.

Seed Packaging

Seed being placed in conditioned storage, which provides a cool, dry environment, will be best stored in packages that allow for air exchange (such as woven poly bags or paper bags).

Storage containers: Seeds stored for extended periods in cold storage, where relative humidity is much harder to control, are better protected when placed in sealed containers such as jars, cans, poly lined bags, or foil packets.

Storage Pests: Consideration should also be given to the effects of seed moisture content, ambient room temperature, and relative humidity in relationship to storage pests. Seed moisture content below 14% typically will not provide conditions that are conducive to the successful development and reproduction of many insects, including most beetles and weevils, and storage mold fungi such as Alternaria, Aspergillus, and Penicillium.These fungi are not capable of growth and reproduction if the seed moisture content is low. Additionally, many insects become inactive at temperatures below 59° F. However, as temperatures increase above 60° F, insect movement, feeding, growth, reproduction, etc., will increase proportionally to the increase in temperature.

Seed Quality

Seed quality is the possession of seed with required genetic and physical purity that is accompanied with physiological soundness and health status. Thompson (1979) defined seed quality as a multiple concept comprising several components and their relative importance in different circumstances and laid much emphasis on

1. Analytical purity / physical purity
2. Species purity / Genetic purity
3. Freedom from weeds
4. Germination percentage
5. Seed vigour and health
6. Seed Moisture content
7. Seed size, weight and specific gravity

Characteristics of a Seed which Decide its Quality

1. **Genetic Purity:** The seed should be true to type. The seed should possess all the genetic qualities / characters, which the breeder has placed in the variety, genetic purity has direct effect on the yields. If there is any deterioration, there would be proportionate decrease in the yield or performance.

Higher Genetically Purity

- Breeder /Nucleus - 100%
- Foundation seed - 99.5%
- Certified seed - 99.0%

2. **Physical Purity:** Physical purity of a seed lot refers to the physical composition of the seed lots. A seed lot is composed of pure seed, inert mater, broken seeds, undersized seeds, soil and dust particles weed seeds, OCS etc. Higher the content of pure seed better would be the seed quality. Pure seed together with germination gives the planting value of the seed lot.

3. **Seed germination and vigour:** Seed germination refers to the ability of a seed when planted under normal sowing conditions to give rise to a normal seedling. Seed vigour refers to the sum total of all seed attributes that give effective plant stand in the field. Higher germination percentage and vigour gives adequate plant population and uniform growth, which have profound effect on yield and determine the planting value of the seed.

4. **Freedom from weeds and other crop seeds:** This is an extension of physical purity described earlier. There are certain weed species, which are very harmful to the crop and once established they are difficult to eradicate. An absolute freedom from seed of such species is highly desirable.
5. **Seed health:** Seed health refers to the presence or absence of disease organisms or insect pests on the seed. The quality of a seed lot depends on its health, hence the seed should be free from seed borne disease and insect pests.
6. **Seed moisture:** The seed moisture is the most important factor in determining the seed germination and viability during storage. At high seed moisture content there is high incidence of pest attack and at moisture content above 16% seed get heated and the viability is lost. Hence the seed should be stored at safe moisture levels of 11-13%
7. **Seed size, weight and specific gravity:** Seed size, weight and specific gravity has been found to have positive correlation with seed germination and vigour in many crops. Therefore the seed should be bold with high specific gravity.
8. **Seed Colour:** The colour of the seed often reflects the condition during seed maturation. The farmers from ancient times have regarded good normal shine as invariable quality guides. The colour and shine deteriorates only when the weather conditions are adverse during maturation or when insects infest the crop or when it is handled badly.

The seed lots having high genetic purity, high germination and with a minimum amount of inert matter, weed seeds and other crop seeds and are free from diseases is said to be of high quality and if it is lacking of these it is said to be of low quality.

Seed Testing

Objectives of seed testing

1. To determine their quality, that is, their suitability for planting
2. To identify seed quality problem and their probable cause
3. To determine the need for drying and processing and specific procedures that should be used.
4. To determine if seed meets established quality standards or labeling specifications.
5. To establish quality and provide a basis for price and consumer discrimination among lots in the market.

Methods used for Testing of seed Viability, Seed Vigour and Genetic Purity of a Variety

1. Seed Viability

Quick Viability Test by Tetrazolium Chloride: The quick viability test is also known as the Tetrazolium Chloride (TZ) test. It's a chemical test that determines seed viability, and it normally takes 24 to 48 hours to get results. The TZ test can provide an early and rapid indication of seed viability, but it is not a substitute for a more thorough seed germination test.

Principle:By hydrogen transfer process catalyzed by cellular dehydrogenases, all respiring tissues may transform a colourless chemical, TZ (2,3,5 triphenyl tetrazolium chloride), to a carmine red coloured water-insoluble formazan. TZ penetrates both live and dead cells, but only living cells catalyse the synthesis of formazan, which is non-diffusible and colours the viable seeds red, but the lack of respiration precludes formazan production, leaving the dead seeds (aged tissue) uncolored.

Germination test: A minimum of 400 seeds are required in four replicates of 100 seeds each or 8 replicates of 50 seeds each or 16 replicates of 25 seeds each depending on the size of seed and size of containers of substrate. Most widely used paper substrates are filter paper, blotter or towel (kraft paper). It should have capillary movement of water, at vertical direction (30 mm rise / min.). It should be free from toxic substances and free from fungi or bacteria. It should hold sufficient moisture during the period of test. The texture should be such that the roots of germinating seedlings will grow on and not into the paper.

2. Seed Vigour

Seedling growth rate: The seedlings arc grown either in laboratory, green house or field. 1n laboratory, in between rolled towel paper method should be followed. Ten seeds are planted in the centre of the moist towel papers in such a way that the micropyles are oriented towards bottom to avoid root twisting. The rolled towel papers are kept in the germinator maintained at a temperature recommended for crop in reference. After a specified period of time (5-10 days) towel papers are removed and five seedlings arc selected, their length is measured and mean seedling length is calculated. Seed lots producing the taller seedlings are considered more vigorous than the seed lots producing shorter seedlings.

Dry weight: For dry weight determination, the seedlings are removed and dried in an air oven at 100°C temperature for 24 hours. The seedling dry weight provides additional information for assessing seed vigour.

3. Genetic Purity

Grow-Out test: The main aim of grow-out test is to determine the genetic purity of the variety of the given sample. In grow-out test plant characters that are less influenced by the environment and which are highly heritable are observed by growing the plants in the field. The variety, which is to be tested for genetic purity, should be grown in the area for which it has been released so that the characters of that variety are fully expressed. Each sample should be sown with proper spacing by adopting the recommended

Electrophoresis

Objective: Verification of variety by electrophoretic mobility of protein on polyacrylamide gel.

Principle: Proteins and enzymes are the primary products of genes and hence are most suited for genetic purity determination. Changes in coding base sequence result in corresponding replacements in amino acids and thus in the primary structure of protein and enzymes. They possess ionizable groups and can therefore be made to exist in solution as electrically charged particles either as cations (+) or anions (-). Molecules with similar charge and size will have differential migration in solution with porous support medium in an electric field based upon differences in net electrical charges as molecules with higher charge migrate faster than those with a lower charge. Particle with smaller molecular weight migrates faster than those with higher weight. This separation of molecules based on their size and net electrical charge is known as electrophoresis cultural practices so that the differences between the varieties are fully expressed.

Safeguards for Maintenance of Genetic Purity of Seed

Maintenance of genetic purity during seed production

Horne (1953) had suggested the following methods for maintenance of genetic purity;

1. Use of approved seed in seed multiplication
2. Inspection of seed fields prior to planting
3. Field inspection and approval of the Crop at critical stages for verification of genetic purity, detection of mixtures, weeds and seed borne diseases.
4. Sampling and sealing of cleaned lots
5. Growing of samples with authentic stocks or Grow -out test

Various steps suggested by Hartman and Kestar (1968) for maintaining genetic purity are as follows:

1. Providing isolation to prevent cross fertilization or mechanical mixtures
2. Roguing of seed fields prior to planting
3. Periodic testing of varieties for genetic purity
4. Grow in adapted areas only to avoid genetic shifts in the variety
5. Certification of seed crops to maintain genetic purity and quality
6. Adopting generation system

Safeguards for Maintenance of Genetic Purity

The important safeguards for maintaining genetic purity during seed production are;

1. Control of seed source
2. Preceding crop requirement
3. Isolation
4. Rouging of seed fields
5. Seed certification
6. Grow out test

1. **Control of Seed Source:** The seed used should be of appropriate class from the approved source for raising a seed crop. There are four classes of seed from breeder seed, which are given and defined by Association of Official Seed Certification agency (AOSCA).
2. **Preceding Crop requirement:** This has been fixed to avoid contamination through volunteer plants and also the soil borne diseases.
3. **Isolation:** Isolation is required to avoid natural crossing with other undesirable types, off types in the fields and mechanical mixtures at the time of sowing, threshing, processing and contamination due to seed borne diseases from nearby fields. Protection from these sources of contamination is necessary for maintaining genetic purity and good quality of seed.

a) **Time isolation:** sowing of seed plot is adjusted in such a manner that the seed plot does not came to flowering at the same time with the neighbouring crop of same variety or other variety of the same crop or related species. Sowing of the seed plot is usually carried 15 to 21 days before or after emergence of the neighbouring crop. This time isolation helps is preventing the genetic contamination of the seed crop as seed plot comes to flowering either before or after the neighbouring crop completes its flowering.

b) **Space Isolation:** Space isolation is the minimum distance kept between the seed plot and neighbouring plot of same crop which prevents natural cross pollination and physical contamination.

Rice (Certified seed production) – 3m

Sunflower (Certified seed production) – 200m

4. **Roguing of Seed Fields:** The existence of off type plants is another source of genetic contamination. Off type plants differing in their characteristics from that of the seed crop are called as off types. Removal of off types is referred to as roguing.

The main sources of off types are

a. Segregation of plants for certain characters or mutations

b. Volunteer plants from previous crops or

c. Accidentally planted seeds of other variety

d. Diseased plants

Off type plants should be rogued out from the seed plots before they shed pollen and pollination occurs. To accomplish this regular supervision of trained personnel is required.

5. **Seed Certification:** Genetic purity in seed productions maintained through a system of seed certification. The main objective of seed certification is to make available seeds of good quality to farmers. To achieve this qualified and trained personnel from SCA carry out field inspections at appropriate stages of crop growth. They also make seed inspection by drawing samples from seed lots after processing. The SCA verifies for both filed and seed standards and the seed lot must confirm to get approval as certified seed.

6. **Grow-out Test:** varieties that are grown for seed production should be periodically tested for genetic purity by conducting GOT to make sure that they are being maintained in true form. GOT test is compulsory for hybrids produced by manual emasculation and pollination and for testing the purity of parental lines used in hybrid seed production.

7. **Cleaning of the instruments**: viz., thresher, sheller, combiner is a must. The machineries should be cleaned prior to each harvesting or threshing and between fields of different parental inbreds or hybrid seed crops. A source of air pressure 700 g/cm2 and a vacuum source are necessary to clean the harvester thoroughly.

3

DNA Finger Printing and Seed Registration

Previous Year Questions

1. What is DNA fingerprinting? Discuss the applied aspects in modern agriculture. (15M, CSE 2021)

- The term DNA fingerprinting in its original sense refers to the method developed in 1985 by Sir Alec Jeffreys (Jeffreys *et al.*, 1985) and his associates for the simultaneous detection of highly variable DNA fragments by hybridisation of specific multilocus probes to electrophoretically separated restriction fragments?
- The DNA fingerprints, resembling barcodes, are unique to the individual and hence can be used in much the same way as conventional fingerprints - to identify individuals with absolute certainty?

DNA Fingerprinting Techniques

- A large number of fingerprinting techniques are in currency, which are variations and combinations of restriction fragmentation and polymerase chain reaction (PCR).
- Restriction fragment length polymorphism (RFLP) involves digesting DNA with restriction enzymes, separating the resultant DNA fragments by gel electrophoresis, blotting the fragments to a filter and hybridising probes to the separated fragments. Specific probe-enzyme combinations give highly reproducible patterns.
- With the advent of PCR technology a number of simpler techniques have become available for molecular characterisation.
- These include, random amplified polymorphic DNA (RAPD) and its several variants, intersimple sequence repeats (ISSR), amplified fragment length polymorphism (AFLP) and sequence tagged microsatellites sites (STMS).

The DNA markers popularly termed as molecular markers belong to four major categories:

1. Southern hybridization based techniques, e.g. RFLP, VNTR
2. PCR based techniques, e.g. AP-PCR, RAPD, DAF, STMS
3. Techniques utilizing basic principles of both RFLP and PCR. e.g. AFLP
4. Technique generating DNA sequence information. e.g. SNP

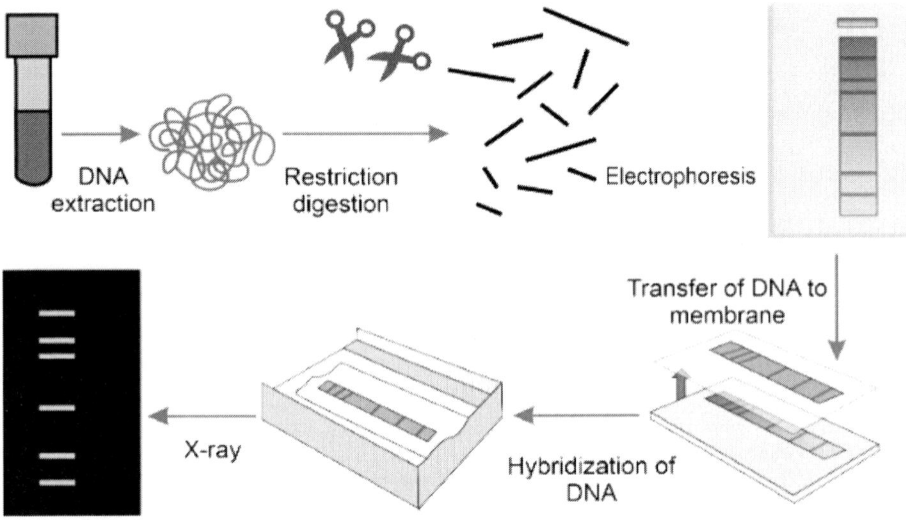

Application of DNA Fingerprinting in Modern Agriculture

Genotype/Cultivar Identification

- Since the last nearly two decades, DNA fingerprinting techniques have found wide application in genotype/cultivar identification in a wide range of crops species comprising cereals and pseudocereals (Echinochloa spp., Hordeum spp., Oryza spp., Secale cereale, Triticum spp., Zea mays), oilseeds (Arachis spp., Brassica spp., Glycine spp.), pulses (Cicer spp., Lens culinaris, Pisum sativum, Phaseolus spp., Vigna spp.), sugar yielding plants (Beta spp., Saccharum spp.), vegetables (Capsicum spp., Cucumis sativus., Lycopersicon esculentum., Solanum spp., Raphanus sativus) and fruits and nuts (Anacardium occidentale, Citrus spp., Mangifera indica, Malus spp., Musa spp., Prunus spp., Pyrus spp., Rubus spp. Vitis spp.).
- The National Research Centre on DNA Fingerprinting has been entrusted with the responsibility of fingerprinting released varieties and important germplasm in crops of national importance.

Patenting of Genes

- The basic data submitted for patenting of genes or their parts are the DNA sequence information.

- While most often patents are sought for gene fragments (cDNA, EST etc.), due to the vagueness of their function and utility their patentability is a wholly contested issue.
- The US Patent and Trademark Office issued new rules of patentability in 2001 that call for "specific and substantial utility that is credible".

Variety Registration and DUS Testing

- In order to be eligible for plant variety registration and the consequent IP protection, a candidate variety must meet the criteria of distinctness, uniformity and stability (DUS).
- Though, morphological data provide the basis for DUS testing, these have some drawbacks.
- Morphological expression is influenced by environment leading to errors in scoring.
- Lack of knowledge of genetic control of phenotypic traits, insufficient variation, and long time required in perennial crops for appearance of the traits at appropriate growth stage are the other limitations that may necessitate use of more reliable and faster to score marker systems.

The advantages of fingerprinting based on molecular markers over morphological character is as include;

i) high degree of non-tissue specific polymorphism,
ii) minimal influence of environment and,
iii) simple inheritance pattern.

- For the present, DNA fingerprint, as proof of unique identity of a plant variety is not acceptable in India or by The International Union for Protection of New Varieties of Plants (UPOV).
- None the less, plant breeders may seek to strengthen their claim for protection of new varieties by including molecular profiles as supplementary information to establish the distinctness of their varieties.
- Molecular profiles may be particularly relevant in cases of biotechnologically developed varieties where only small apparent phenotypic differences exist between new variety and an extant one.

Resolving Breach of Intellectual Property and Biodiversity Protection

- DNA fingerprinting can be used to provide proof of or defense against allegations of breach of intellectual property rights.
- Such infringements would occur when

i) a registered variety is cultivated/ marketed unauthorisidely under its own or a different name,

ii) plant material comprising seeds, flowers, fruits or other plant products are falsely sold under the name of a registered variety,

iii) plant material is collected from the wild and commercially exploited without the authorization of biodiversity authority.

- Proof of infringement in all the above cases would require a rapid and unambiguous method of identification.
- Detection of infringement of breeders' rights in fruit trees by conventional DUS tests would be impossible since it would require nursery trees and other seedlings to bear fruit for comparison, a delay of several years.
- Use of DNA fingerprinting and similar other methods is, therefore, highly recommended for identification of tree varieties. Below are given some examples where DNA fingerprinting has actually aided in protection of intellectual property and related aspects in crops.

DNA fingerprinting for Essentially Derived Varieties (EDV)

- On the flip-side, IP protection may trigger disputes and therefore means are proposed to manage and restrict commercialisation of close copies or plagiarised versions of an already existing variety through the concept of EDVs.
- EDVs are expected more from a biotech industry armed with genes and transformation protocols. Unless provided with an appropriate environment, traits such as resistance are difficult to identify using morpho-agronomic descriptors.
- DNA markers will be useful in such situations. If the transgene is mandated to be disclosed, DNA fingerprinting can quickly identify the variety and settle the issue.

4

Role of Public and Private Sectors in Seed Production and Marketing

Previous year questions

1. Introduction?
2. Role of private?

Introduction

- The Indian seed industry used to be dominated by public sector seed companies. However, following the easing of government regulations and the implementation of a new seed policy in 1988, the private sector seed companies have started playing a major role in seed development and marketing.
- More recently, the government's decision to embrace biotechnology as a means of achieving food security has attracted several leading biotechnology-focused multinational seed companies to India.
- The composition of the seed industry, by volume of turnover, has reportedly reached a ratio of 60:40 between the private and public sectors.

Public Sector Seeds Companies

National Seed Corporation (NSC)

- Public sector involvement in the seed industry on a national scale began at the beginning of the "green revolution" with the establishment of the National Seed Corporation (NSC) in 1963, which was charged with the responsibility of promoting seed industry development from production through processing, storage and marketing, and establishing a system of quality control.
- Before that, the Indian seed industry was little developed apart from a small number of private companies dealing with high value vegetable and flower seeds.

- In the initial years of operation, the NSC concerned itself mainly with foundation seed production and with seed certification after the enactment of Seed Act in 1966.

States Farms Corporation of India Limited (SFCL)

- It was formed in 1969 with the mandate of producing breeder, foundation, and certified seeds of high yielding varieties. With its 12 large mechanized farms located in 8 different states, the SFCL continues to be the largest seed-producing agency in the country.

State Seed Corporations (SSC)

- The State Seed Corporations (SSC) were established later with support from the World Bank, initially in nine states, and later expanded to cover 13 states, for production and handling of seed in their respective states.

Role of Public Sector Seed Companies

- It is now mostly confined to certified seeds of high volume, low value segment of high yielding varieties of cereals, pulses, and cotton with a limited presence in the high value hybrid sectors of cotton, cereals, and vegetables.
- Wheat and paddy seed constitutes a major share of the seeds handled by them. The NSC and SSCs work closely together to coordinate procurement and sales prices as well as variety demand and supply.
- Their presence is considered necessary by the government to ensure the availability of reasonably priced seeds of major crops throughout the country and to make sure that private sector seed companies do not enjoy and exploit unreasonable market power.
- The public sector seed companies, however, lag behind in research; they are mostly dependent on public research institutions, under the aegis of Indian Council of Agricultural Research (ICAR) and State Agricultural Universities (SAUs) for their breeder seed requirements.
- Based on feedback from dealers and end-users, the public sector seed companies/state governments forecast seed demand for various crops three years in advance and a requirement for breeder seeds is placed with the GOI's Ministry of Agriculture.
- Using the breeder seeds supplied by government research institutes, the public sector seed companies produce foundation seeds on government farms or reliable, well-trained contract farms. These are further multiplied in contract farmers' fields next year as certified seeds for commercial distribution.

- If for some reasons (drought or other weather calamities) the supply of certified seeds falls short of requirements, the public sector seed companies source commercial grain from the market, upgrade the quality, and after proper testing distribute it as quality seeds.
- All seed grown by contract growers for seed corporations meeting the specified standards attract a premium price over and above the commercial grain price for that crop. The premium can vary between 25 percent for cereals to over 100 percent for hybrids.
- In the public sector, NSC is usually the retail price setter with the SSCs following NSC prices in determining their own for similar or substitute varieties.
- For selfpollinated field crops, an accepted basis is to add a margin of 15 to 25 percent on production costs. For hybrid seeds of cereals and vegetables, prices to some degree reflect market trends.
- However, there is government intervention in the pricing of seeds produced by public sector corporations with the degree of intervention varying from state to state.
- Some states are now thinking of giving greater autonomy to their seed corporations to make them financially viable by allowing them to market private branded seeds, domestically produced or imported.
- An advantage to the government seed companies is that they have a vast distribution network and trusted brand image. The reason why they are losing market share is because seeds by private companies often outperform the publicly available varieties.
- Some SSCs have started their own research to evolve superior propriety hybrids.

Private Sector Seed Companies

Most large multinational seed companies now have their presence in India (either as a joint venture or with 100 percent equity) with their main focus on biotechnology. These include

- Monsanto,
- Bayer CropScience,
- Syngenta, Advanta,
- Hicks-Muse-Tate,
- Emergent Genetics,
- Dow Agro,

- Bioseed Genetics International Inc.,
- Tokita Seed Co, and
- Nunhems Zaden BV.
- Easing of government regulations in the late 1980s spurred enormous development within the seed industry by attracting several foreign seed companies to India.
- While some of them (like Cargill) entered through joint venture partnerships with Indian seed companies, some others already had a presence in India through affiliate companies (like Hindustan Lever).

Role of Private Sector Seed Companies

- They identified potential crops for hybridization and started research and development activities. Typically they concentrated on hybrids, mainly corn, cotton, sunflower, vegetables, and flowers (more recently on rice) and they now account for a major share of commercial production of these seeds in India.
- The basic reason for the private sector's focus on these crops is that it involves low production volume and higher margins.
- Which involve high volume and low margin and are more prone to piracy in the absence of an effective Plant Variety Protection regime in India.
- Furthermore, there is no significant government intervention in the pricing of these hybrids, and the Indian seed regulations permit marketing of truthfully labeled seeds.
- Currently, some 500 hybrids of field crops and vegetables are being marketed, as truthfully labeled seeds, mostly by private seed companies.
- The private seed sector now comprises some twenty or so large players (with sales turnover exceeding rs. 200 million), several medium companies (sales turn over between rs. 200 million and 20 million), and a large number of small, unorganized players (sales turnover less than rs. 20 million) with local presence.
- The private seed industry is now undergoing a transition following the Indian government's focus on biotechnology research, as a means of increasing agricultural production and also driven by trends in the domestic and world seed market.
- Intensifying international competition, increasing R&D costs, and the complexity of biotechnology have lead to increased consolidation of the Indian seed industry with several of the large and medium companies merging or being taken over by multinational seed companies.

- Another factor attracting international seed companies to India is the country's varied agroclimatic conditions and abundant skilled and unskilled labor, as seed production, particularly hybrid seed production, is highly labor intensive.
- Private seed production is largely centered around Bangalore for vegetable crops and Hyderabad for field crops, particularly cottonseeds.
- The initial focus of many of these companies has been cottonseed, for which genetically modified (Bt) hybrids have already been approved by the Indian government for commercial cultivation, with other bio-engineered crops in the pipeline.
- Most of these companies have licensing agreement with Monsanto for the Bt gene; some are trying to develop their own Bt technology, legally or illegally.
- The seed industry is represented at the national level by two associations **"The Seed Association of India"** based in **New Delhi** and the **"Association of Seed Industries"** based in **Mumbai**. Recently, a third association called **All India Crop Biotechnology Association (AICBA)**, was formed with members from mostly hybrid seed producers and multinationals like Monsanto and Dow Chemicals.

Public-Private Sector Cooperation

- Cooperation between private sector seed companies and public research institutes under ICAR, SAUs, and the International Crop Research Institute for Semi-Arid Tropics (ICRISAT), supported by the Consultative Group on International Agricultural Research (CGIAR), is growing.
- Public sector breeder seeds are available free of charge to private seed companies with no strings attached. The AICT annual workshops provide venues to private sector seed companies to assess what is available with public research institutes.
- Under the "consortium" model with ICRISAT, private companies can jointly fund research that results in publicly available parental lines, which they often cross with in-house genetics to produce proprietary hybrids.
- ICRISAT recently introduced a live-in campus for private sector researchers to use the institutions' facilities and expertise. ICRISAT is focusing more on private sector partnerships for funding reasons and also because of private companies' effectiveness in getting the research result out to farmers.

- ICRISAT is currently reviewing its policy of keeping all research in the public domain and is considering licensing/royalties/exclusive rights.
- Private companies can also fully fund research at SAUs for exclusive rights on the results and/or hire professors as consultants, although the degree of cooperation varies from state to state.

Role of Seed Industry in the Development of Indian Agriculture

- The growth of seed industry can be considered parallel to the growth of India's agriculture production.
- Today, the Indian seed industry is the fifth largest seed market in the world, accounting for 4.4% of global seed market after the U.S. (27%), China (20%), France (8%) and Brazil (6%).
- From a predominently public sector producing varieties suited for Green Revolution agriculture in the 1960's, the Indian seed industry has evolved into a multi-faceted industry with a large involvment of private firms and increasing emphasis on research and development.
- The main task of the public sector today lies in meeting the national seed requirement for high volume, low value crops like self-pollinated crops such as rice or wheat.
- These seeds are supplied at fixed rates (determined by the government), and the NSC and SSC have incurred major deficits, because fixed prices do not always reflect the actual costs of production, processing, and distribution.
- Private industry developed their own hybrid varieties of crops such as pearl millet, sunflower, cotton, sorghum, and maize. These private hybrids represent an important share of the market for these crops and resulted in drastic increase in production.
- The production of hybrid seeds is to a large extent the prerogative of the private sector. Of all improved cutivars bred and marketed by private companies until 1993, nearly 70% were hybrids.
- Since hybrid seeds cannot be multiplied in farmers' fields, they must be bought from the company every time they are raised.
- Apart from hybrids, the private sector is also largely involved in the commercialization of low volume, high value crops such as vegetable seeds.
- With some 500 companies of various sizes, 24 of which with links to multinationals, the private sector contributes a little less than 50% of the commercial seed requirement for the country today.

- Over the period of time the SRR of various crops has been increasing from 10% to more than 50% in many of the crops where Open Pollinated Varieties (OPV) seeds are used. This leads to a large section of farmers using Farmer saved seed and being out of purview of the commercial seed industry.
- It should be noted that in case of crops where hybrid seeds are used, the SRR is 100% as the farmers cannot re-use the seed for the next generation as re-use of hybrid seed leads to dilution of genetic purity due to genetic segregation of characters in the next generations.
- This high seed replacement rate ensures firms good sales. That is not the case with high yielding varieties of crops like wheat or rice, for which the replacement rate fluctuates between 9 and 14%.
- While certain large companies have competencies in all the industry functions such as R&D, Production, Processing, marketing and distribution, small companies specialize in one or more functions.
- The public sector research of State Agricultural Universities (SAUs) and Indian Council of Agricultural Research (ICAR) has not only fuelled the development of Indian seed industry, but also continues to drive the industry with continuous delivery of new improved plant varieties of Plant varieties to both Public and private sector.

Transgenics

By 2010, 85% of Indian Cotton was converted into GM Cotton and presently nearly 99% of Cotton grown in India is GM Cotton. Bt Cotton is the only GM crop allowed for cultivation in India. Presently India has the fifth largest GM cropped area in the world and the GM Cotton market in India is nearly 16% of the total commercial seed market in India.

- The entire hybrid Cotton seed market is GM Cotton seed and therefore GM seed market is nearly 16% of the total seed market and 35% of the hybrid seed market.

5

Intellectual Property Rights (IPR) Issues, WTO Issues and Its Impact on Agriculture

Previous Year Questions

1. Intellectual property rights (IPR) are the rights given to persons over the creations of their minds: inventions, literary and artistic works, and symbols, names and images used in commerce. They usually give the creator an exclusive right over the use of his/her creation for a certain period of time?
2. Plant varietal protection?

- The importance of intellectual property was first recognized in the **Paris Convention for the Protection of Industrial Property (1883)** and the **Berne Convention for the Protection of Literary and Artistic Works (1886)**. Both treaties are administered by the **World Intellectual Property Organization (WIPO)**.

1. **Trade secret:** When the individual / organization owning an intellectual property does not disclose the property to any one and keeps it as a closely guarded secret to promote his business interests, it is called trade secrets. Trade secret may relate to formulae, processesor parented lines in hybrids, in biotechnology trade secret include cell lines, micro organism strains etc.

Example: The secret formula for Coca-Cola, which is locked in a vault

2 **Patents:** A Patent is the right granted by a government to an inventor to exclude others from imitating, manufacturing, using or selling the invention in question for commercial use during the specified period.

Patent Requirement:

1. Novelty
2. Inventiveness
3. Industrial application & usefulness
4. Patentability
5. Disclosure

Novelty: The invention must be new and should not be already known to public.

Inventiveness: The invention should represent an innovation

Industrial Application & Usefulness: The patent must have an industrial application should be useful to the society/nation.

Patentability: It must be patentable under the existing law and its current interpretation.

- The Indian Patent Act 1970 does not allow product patents in pharmaceuticals, food and agriculture.
- The key element is that substances used as food/medicine/drug and the entire class of materials formed by any chemical reaction do not qualify as patentable subject matter i.e. product patents are not allowed in India.
- As a result, an antibiotic is not patentable in India, while the process of its production is in contrast; both the product & the processes are patentable in Europe & USA.

Disclosure

- The inventor has to describe his invention in sufficient detail so that a person of normal skill is able to reproduce it.
- In case of biological entities already known, organisms may be simply named. But if they have been genetically modified, the nature and the method of modification has to be described fully.
- The disclose of an invention gives an opportunity to other inventors to improve upon the various features of the invention, so that it became more efficient & /or more useful.
- This in-turn, results in scientific and economic progress of the society/ nation.

Limits of a patent: A patent is limited both in time and space

a) **Limitation of Time** – A patent is valid for a specified period of time from the date of award in most countries this period is 15-20 years. The Indian patent act (1970) grants protects for 7 or 14 years. However, there is a strong argument for larger protection as it ay take upto 10 years from the time or patent is awarded to the time the product reaches market.

b) **Limitation of Space** – A patent is valid only in the country of its Award and not in other countries. A group of nations may agree to honour the patents awarded by any member country eg. European Economic community. WTO has a similar provision that a patent awarded by WTO will be valid in all member countries.

3. **Copyright Copyright** is an intellectual property right that law gives to a creator of literary, dramatic, musical, and artistic work and a producer of cinematograph films and sound recordings. It also applies to architectural works and computer program/software. It can be understood as a bundle of rights that include the right of reproduction, communication, adaptation, and translation of the work. Copyright ensures protection to the rights of authors over their creations and in turn aims at rewarding creativity.
4. **Geographical Indication (GI)**: a sign used on a product that originates from a specific geographical location. The product must possess reputation and qualities of the place of origin. GI are generally registered on products produced by rural, marginal and indigenous communities over generations that have garnered massive reputation at the international and national level due to some of its unique qualities. GI tag gives the right to only those registered users the right to use the product name, and prevents others from using the product name that does not meet the standards prescribed.

GI tags are used on the following types of products.

- Handicrafts – Examples would be Madhubani Paintings, Mysore Silk
- Food items – Example would be Tirupati Laddu, Rasgulla.
- Wine & Drinks – Example would be Champagne, Cognac of France; Scotch Whisky of UK, Tequila of Mexico.
- Industrial Products
- Agricultural Products – Basmati Rice

5. Trademark

- A trademark is a sign capable of distinguishing the goods or services of one enterprise from those of other enterprises.
- In India, trademarks are governed by the Trade Marks Act 1999, which was amended in 2010.
- It legally differentiates a product or service from all others of its kind and recognizes the source company's ownership of the brand.
- Although trademarks do not expire, the owner must make regular use of it in order to receive the protections associated with them.
- It serves as a badge of origin exclusively identifying a particular business as a source of goods or services.
- Trademark infringement is the unauthorised usage of a sign that is identical or deceptively similar to a registered trademark.

6. Industrial Design

- An industrial design renders an object attractive or appealing, thus increasing its marketability and adding to its commercial value.
- The design may be three-dimensional based on the shape or surface of the object, or two-dimensional based on the object's patterns, lines or colours.
- Legally, "industrial design" is the title granted by an official authority, generally the Patent Office, to protect the aesthetic or ornamental aspect of an object.
- Industrial design rights are granted to the creator of designs to reward them for their effort and investment in manufacturing the product. These rights enable the owner to make articles to which the design is applied or in which the design is embodied.

7. Plant Varietal Protection

Protection of Plant Varieties and Farmers' Rights Act, 2001

- In order to provide for the establishment of an effective system for the protection of plant varieties, the rights of farmers and plant breeders and to encourage the development of new varieties of plants it has been considered necessary to recognize and to protect the rights of the farmers in respect of their contributions made at any time in conserving, improving and making available plant genetic resources for the development of new plant varieties.
- The Govt. of India enacted "The Protection of Plant Varieties and Farmers' Rights (PPV&FR) Act, 2001" adopting sui generis system. Indian legislation is not only in conformity with International Union for the Protection of New Varieties of Plants (UPOV), 1978, but also have sufficient provisions to protect the interests of public sector breeding institutions and the farmers.
- The legislation recognizes the contributions of both commercial plant breeders and farmers in plant breeding activity.

Objectives of the PPV & FR Act, 2001

- To establish an effective system for the protection of plant varieties, the rights of farmers and plant breeders and to encourage the development of new varieties of plants.
- To recognize and protect the rights of farmers in respect of their contributions made at any time in conserving, improving and making

available plant genetic resources for the development of new plant varieties.

- To accelerate agricultural development in the country, protect plant breeders' rights; stimulate investment for research and development both in public & private sector for the development new of plant varieties.
- Facilitate the growth of seed industry in the country which will ensure the availability of high quality seeds and planting material to the farmers.

Criteria for Protection

A new plant variety shall be protected by registration if it is novel, distinct, uniform and stable.

1. **Novel:** A new variety shall be deemed to be novel. If at the date of filing of application for registration for protection, the propagating or harvested material of such variety has not been sold or otherwise disposed of by or with the consent of the breeder, for the purposes of exploitation of such variety
   a) in India, for more that one year or
   b) outside India in the case of trees or vines, earlier than 6 years or, in any other case, earlier than four years.
2. **Distinct :** A new variety shall be deemed to be distinct if it is clearly distinguishable by at least one essential characteristic from any commonly known variety.
3. **Uniform:** The new variety shall be deemed to be uniform if, subject to the variation that may be expected from the particular features of its propagation, it is sufficiently uniform in its essential characteristics
4. **Stable:** A new variety shall be deemed to be stable if its essential characteristics remain unchanged after repeated propagation or, in the case of a particular cycle of propagation, at the end of each such cycle. The requirement of stability means that a variety must remain true to its description/character even after repeated propagation.

Rights under the Act

Breeders' Rights

- Breeders will have exclusive rights to produce, sell, market, distribute, import or export the protected variety.
- Breeder can appoint agent/ licensee and may exercise for civil remedy in case of infringement of rights.

Researchers' Rights

- Researcher can use any of the registered variety under the Act for conducting experiment or research.
- This includes the use of a variety as an initial source of variety for the purpose of developing another variety but repeated use needs prior permission of the registered breeder.

Farmers' Rights

- A farmer who has evolved or developed a new variety is entitled for registration and protection in like manner as a breeder of a variety;
- Farmers variety can also be registered as an extant variety;
- A farmer can save, use, sow, re-sow, exchange, share or sell his farm produce including seed of a variety protected under the PPV&FR Act, 2001 in the same manner as he was entitled before the coming into force of this Act provided farmer shall not be entitled to sell branded seed of a variety protected under the PPV&FR Act, 2001;
- Farmers are eligible for recognition and rewards for the conservation of Plant Genetic Resources of land races and wild relatives of economic plants;
- There is also a provision for compensation to the farmers for non-performance of variety under Section 39 (2) of the Act, 2001 and
- Farmer shall not be liable to pay any fee in any proceeding before the Authority or Registrar or the Tribunal or the High Court under the Act.

Rights of Indigenous Community

- It is compensation to village or local communities for their significant contribution in the evolution of variety which has been registered under the Act.
- Any person/group of persons/governmental or non- governmental organization, on behalf of any village/local community in India, can file in any notified centre, claim for contribution in the evolution of any variety.

8. Protection of Traditional Knowledge

Traditional knowledge generally refers to the knowledge system that has been long embedded in the culture or tradition of indigenous, regional, or local communities. Culture, biological resources, and knowledge in traditional societies are closely interlinked and interwoven to form an intricate part of their heritage as well as provide a distinctive identity to the society. These kinds of knowledge are crucial for such communities' existence and survival.

Need for Protection of Traditional Knowledge

- It seeks to prevent **misappropriation, misuse and unauthorised use** of TK
- Control of use of TK beyond the traditional and customary context, **fair and equitable benefit sharing**, and encouragement of tradition based innovation.
- It seeks to have an instrument that contributes to protection of innovation and to the transfer and dissemination of knowledge to the **mutual benefit of holders and users of protected TK in a manner conducive** to the social and economic welfare and to balance the rights and obligations.
- There has been an **increased interest in traditional knowledge for commercial use** by third parties especially as it relates to genetic resources & traditional cultural expressions for trade, **pharmaceuticals**, agriculture etc.
- TK is usually passed down from generation to generation & until recently was not captured in writing. This creates a **risk of knowledge being lost when the holder dies**, without documentation.
- Protection against **unfair competition** between developed and developing countries.
- If it remains unprotected, **indigenous people** who are responsible for bringing it to the forefront can **end up suffering huge losses**, on an emotional as well as a financial level.

Sui Generis Protection of Traditional Knowledge

- Sui generis is defined as something "of its own kind" that is special, unique, and different. Conventional forms of I.P.R are not sufficient to protect indigenous knowledge since they are based on the protection of individual property rights.
- In contrast, Traditional knowledge (TK) and Traditional Cultural Expression (TCE) call for a larger collective because it involves communities.
- The aim of incorporating a sui generis system in India for the protection of T.K and T.C.E could be done by modifying certain features of the existing intellectual property regime in order to accommodate the unique characteristics of its subject matter, i.e. traditional knowledge and traditional cultural expressions along with understanding the policy requirements, beneficiaries and examining the concept of equitable sharing and the principle of prior informed consent.

- Such a system can help legal rights associated with T.K and T.C.E and thereby create space for access and benefit-sharing to protect the diverse set of knowledge and information that Indian indigenous communities hold.
- While drafting a sui generis system for the protection of traditional knowledge and traditional cultural expressions, it is important to analyze certain elements such as:
  - Disclosure of the genetic and biological resources of the country of origin, preferably sui generis database protection system which will ensure accountability and tracking
  - Provisions of prior informed consent and equitable benefit-sharing between the rightful custodian holders of such traditional knowledge and expressions and the third parties.
  - Preservation of cultural identity and protection of communities from being threatened
  - Understanding ownership and distinctiveness as well as the novelty of such traditional knowledge and expressions
  - Provisions of customary international law
  - Disclosure of existing traditional knowledge and expressions
  - Issues of biopiracy and conservation of the environment
  - Promotion of access and benefit-sharing

The kinds of intellectual property protection that can be incorporated in the sui generis system can be of two kinds-

1. **Defensive protection**: wherein people from outside the community holding such traditional knowledge will be prohibited from gaining intellectual property rights over the subject matter. This will safeguard the interests of the rightful custodian holders and prevent undue misuse. For example, India has a repository containing a database of all the traditional medicines that can be used as evidence for prior art when examining patent applications. Such an incident gave rise to a famous case wherein the US Patent and Trademark Office allowed registration of a patent for the use of the spice turmeric that could be used to treat bruises; the subject matter of treatment being turmeric has been long known to Indian traditional and regional communities The patent was later revoked.
2. **Positive Protection**: This type of protection seeks to grant legal rights to help and accredit local and regional communities to promote their

traditional cultural expressions and knowledge, have it in their power to control its uses and utilization, and most importantly, control its benefits from the possible commercial exploitation.

Traditional Knowledge Digital Library (TKDL): is a database of over 2,50,000 formulations used in traditional medicine systems in India, namely, Ayuveda, Siddha, Unani and Yoga.

- TKDL is a pioneer initiative of India to prevent misappropriation of country's traditional medicinal knowledge at international patent offices on which healthcare needs of more than 70% population and livelihood of millions of people in India is dependent.

WTO Issues:

- The World Trade Organisation's Agreement on Trade-Related Aspects of Intellectual Property Rights (TRIPS) is an international legal agreement between all WTO members (WTO). It defines minimum requirements for national governments to regulate various forms of Intellectual Property (IP) as they apply to citizens of other WTO member countries.

Trade-Related Aspects of Intellectual Property Rights (TRIPS)

- TRIPS was negotiated between 1989 and 1990 at the close of the Uruguay Round of the General Agreement on Tariffs and Trade (GATT) and is administered by the World Trade Organisation (WTO).
- TRIPS defines minimum requirements for trademarks, copyrights, geographical indications, patents, industrial designs, layout designs for integrated circuits, and concealed information or trade secrets in terms of availability, scope, and usage.
- It applies to member states core international trade norms relating to intellectual property.
- It is applicable to all members of the WTO.
- WTO members have the freedom to pursue their domestic policy goals.
- It places the IP system in the context of innovation, technological transfer and the public good.
- The TRIPS Council is in charge of implementing and overseeing the TRIPS Agreement's operation.
- TRIPS is sometimes known as a "Berne and Paris-plus" agreement.

TRIPS Agreement – Doha Declaration

- In 2001, the Doha Declaration was ratified by WTO members. It reiterated TRIPS' adaptability in situations incorporating patent rights.

- This was done to facilitate access to appropriate medications.
- According to the Doha Declaration, the TRIPS agreement will not preclude governments from undertaking initiatives to protect public health.
- This loosened TRIPS guidelines. Everyone has recourse to life-saving medicines and immunizations. Pharmaceutical firms must not benefit from it.